YOUR KNOWLEDGE HAS VALUE

- We will publish your bachelor's and master's thesis, essays and papers

- Your own eBook and book - sold worldwide in all relevant shops

- Earn money with each sale

Upload your text at www.GRIN.com and publish for free

Bibliographic information published by the German National Library:

The German National Library lists this publication in the National Bibliography; detailed bibliographic data are available on the Internet at http://dnb.dnb.de .

Imprint:

Copyright © 2014 GRIN Verlag, Open Publishing GmbH
Print and binding: Books on Demand GmbH, Norderstedt Germany
ISBN: 9783668257955

Gunther Klobe

Aus der Reihe: e-fellows.net stipendiaten-wissen

e-fellows.net (Hrsg.)

Band 1994

Reverse bias and PIN diode sensitivity. Building a solid state radiation detector

GRIN Publishing

Reverse bias and PIN diode sensitivity

Gunther KLOBE

Summer of 2014

Abstract

A prototype of a solid state radiation detector based on a PIN diode has been developed to the point where it was ready for a test flight on a weather balloon. This device is intended to be used to monitor secondary cosmic radiation and may have the potential to be produced commercially since space weather is an issue of growing importance concerning the economy and technology of modern civilisation. Some of the most crucial experimental work in preparation for the test flight is discussed in this report. One important question was what reverse bias should be applied to the PIN diode. As a result, it was found that the detector's sensitivity increases with increasing reverse bias until it reaches a saturation value at $\sim 25\text{V}$. Another concern was that the prototype may not be robust against temperature differences. This concern could be ruled out to a certain extent.

Contents

1 Introduction

In an ongoing project in the Department of Physics at the University of Oxford there's an effort to build a solid state detector for atmospheric ionisation measurements, in particular the measurement of secondary muons generated by cosmic rays.

During a 7 week project, several tasks related to the development and laboratory testing of the device were completed, such that in the end the detector was ready for a test flight on a weather balloon. The most important tasks will be discussed in this report.

2 Motivation

Galactic cosmic radiation (GCR) has effects in the atmosphere that are still not understood very well.

Also, space weather is dictated by GCR and it is very important for the safety of spacecrafts, especially satellites. It has been estimated that during the period 1994-1999 alone, over 500 000 000 $ in insurance claims were disbursed due to on-orbit failures of satellites related to space weather. [1]

It should come as no surprise that space weather is of concern both scientifically and commercially, especially in this day and age where there are more satellites than ever before orbiting Earth and there is even more money at stake than in the aforementioned period.

Still, there's a discrepancy between models and measurements that needs to be understood better. Further research in this field could be facilitated with the Small Radiation Detector because of its unique qualities compared to other devices used in ionisation measurements. Typical measurements can be done with different kinds of technology: Ionisation chambers, Geiger counters or solid state detectors.

Ionisation chambers contain an isolated electrode in a chamber filled with a fixed amount of gas. The ionisation rate can then be calculated from the rate of decay of voltage on said electrode. Cosmic rays were first discovered in 1908 by Bergwitz using this technology. He didn't publish his results, though, because he suspected an error in his measurements. However, he told his colleague Victor F. Hess about it and in 1912 the latter could prove in a more rigorous experiment the existence of what is now called secondary cosmic radiation. [2]

With Geiger counters, the idea is to have a voltage of a several hundred volts across a low pressure noble gas in a sealed pipe. When any ionising particle enters the pipe, the gas gets ionised in a cascade effect which results in a drop in voltage that can be measured and thus the amount of such events can be counted. While ionisation chambers give us some idea of the energy distribution of ionising particles from looking at the length of the track that such a particle leaves in the chamber, Geiger counters don't have that feature. Yet, Geiger counters are widely used for air ionisation measurements because the technology is simple and effective compared to ionisation chambers. However,

Geiger counters have further disadvantages, including fragility and the need for high voltage, which can be difficult or expensive to provide on a test fligth.

In a PIN diode setup under reverse bias, energetic particles can ionise solids just like they ionise gas in the two measurement methods discussed above. However, PIN diodes enable energy discrimination and can be run at very low voltages compared to Geiger counters ($\sim 10V$ instead of $\sim 500V$). This is why a solid state detector can be a good alternative for these kinds of measurements.

3 Small Radiation Detector set-up

Figure 1 shows the basic set-up of a PIN photodetector. Basically, it is a PIN diode with a voltage applied such that the negative terminal is connected to the p-doped material and the positive terminal is connected to the n-doped material. The result of this so-called reverse bias is that holes (positive charge carriers) in the p-type region and electrons (negative charge carriers) in the n-type region get pulled away from the depletion layer. This means that the depletion layer widens and in a static situation there is practically no current flowing, apart from a small residual, "dark" current I_S. Furthermore, the I-layer between the p- and the n-layer is an intrinsic semiconductor, which also increases the size of the depletion area by a multitude. This is important in a photodetector because only electron-hole pairs created by incident photons in (or near) the depletion area will get swept out of the region by the reverse bias which then creates a current. In conclusion, a large depletion region is needed for maximum efficiency of the photodetector. As one can see from figure 1, only photons (or other kinds of ionising radiation) of sufficient energy, i.e. energy higher than the band gap in the intrinsic semiconductor, will create electron-hole pairs in the active area of the PIN diode.

The PIN diode used as a detector in this project is the PS100-7-CER-2. Because we are mainly interested in high energy radiation, namely muons, the sensor circuit needs to be shielded from lower energy radiation, like visible light, which would result in unnecessary noise. This is done by wrapping the sensor circuit in copper foil, as shown in figure 2.

Figure 3 shows the complete set-up of the Small Radiation Detector. The parcel in figure 2 is installed on a PCB (printed circuit board) with different PICs. Its main functions are to amplify the signal coming from the photodiode and to convert these analogue pulses into digital ones.

The whole detector runs on a supply voltage of 16V, has dimensions of approximately $10cm \times 5cm \times 1cm$ and weighs about 30g. The data output is RS232/TTL to USB. It gives data in the ASCII format about the total number of counts, the time since start in seconds and milliseconds, as well as the pulse height.

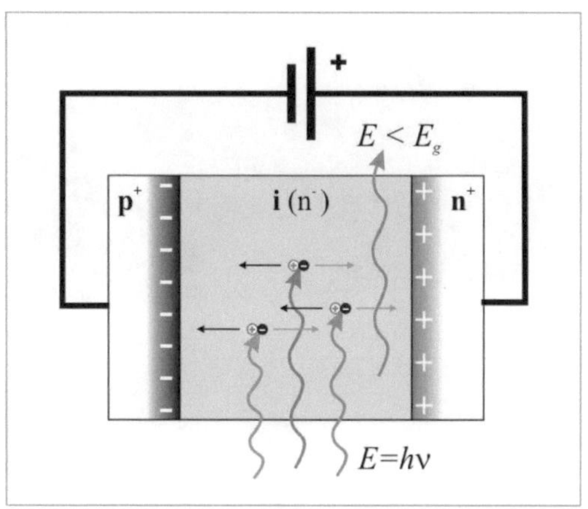

Figure 1: Schematic of a PIN photodiode with reverse bias, credit to [3]

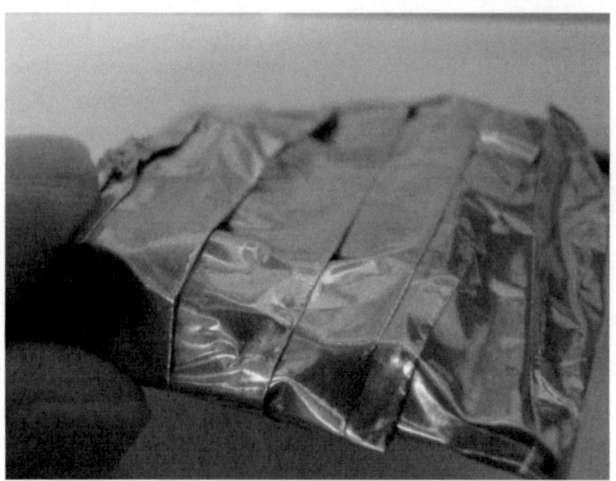

Figure 2: Sensor circuit, including the PIN diode, wrapped in copper foil

Figure 3: Complete detector: Parcel on PCB with PIC microcontrollers

4 Laboratory testing

4.1 Finding the optimal reverse bias voltage

4.1.1 Introduction

The sensitivity of the PIN photodiode used in the Small Radiation Detector project, PS100-7-CER-2, was tested by measuring signal rates for different reverse bias voltages while exposing the photodiode to a radioactive source of a known, constant amount of radiation.

The active region in the PIN photodiode is identical to the depletion layer within the intrinsic semiconductor (the I-part of the PIN-photodiode). The size of the depletion layer increases when the reverse bias voltage is increased, up to a point where the entire intrinsic semiconductor is depleted (saturation).

In this experiment, the goal was to find out how the depletion layer size is scaling with reverse bias, and in particular to find the saturation point for the reverse bias to allow for maximum sensitivity. It is important to know this value because applying reverse bias voltages higher than this would be unnecessary and a waste of energy.

4.1.2 Experiment

In order to learn more about the effect of different reverse biases on the sensitivity of the PIN photodiode, the parcel containing it was separated from the rest of the Small Radiation Detector. Cables were attached such that the reverse bias could be changed and regulated by a DC power supply.

The measurements at different reverse bias voltages were done using an amplifier (ORTEC 440A), adjusted for a gain of 32 and a pulse shaping time of $0.25\mu s$

and a single channel analyzer (SCA; Canberra 2030). A radioactive source (2U435 from the Oxford Physics Teaching Labs) was placed closely above the parcel. This is a Co60 source which emits gamma radiation of either 1173.24keV or 1332.5keV. Its activity is $\sim$ 14kBq. In order to reduce noise as much as possible, the lower-level discriminator (LLD) in the SCA was set to LLD = 400mV. In order not to cut off any signal, the upper-level discriminator was set to the maximum of ULD = LLD + 10V = 10.4V. These numbers don't correspond to the pulses inside the PIN diode. They correspond to the signal *after* going through the operational amplifier inside the parcel *and* the ORTEC 440a amplifier mentioned above.

In a first series of experiments, the number of pulses in this setup was counted (using the SCA) in 5min runs for different reverse bias voltages ranging from 0V to 45V. In a second series, the radioactive source was removed and then the same measurements as before were performed in order to identify the background for the respective reverse bias voltages.

4.1.3   Results and data analysis

The results of these measurements, converted into the corresponding count rates, are listed in the second and third column of the table below.

Computing the difference between the count rate and the background is a straightforward way of finding the signal rate, which can be found in the last column of the table. The uncertainty in the values of the signal rate were calculated assuming a Poisson distribution of events, as usual when it comes to radioactive decay.

Voltage [V]	count rate [Hz]	background [Hz]	signal rate [Hz]
0.00	0	0	0
5.00	0.19	0.01	0.18 ± 0.02
10.00	1.31	0.01	1.30 ± 0.07
15.00	1.81	0.02	1.79 ± 0.08
20.00	2.37	0.03	2.34 ± 0.09
25.00	3.02	0.01	3.01 ± 0.10
30.00	2.95	0.04	2.91 ± 0.10
35.00	3.17	0.01	3.16 ± 0.10
40.00	3.06	0.05	3.01 ± 0.10
45.00	3.14	0.06	3.08 ± 0.10

Figure 4 shows the correlation between reverse bias voltage and signal rate according to the values above.

As one can see from the plot, the signal rate correlates almost linearly with the reverse bias until it reaches a saturation value (of $\sim$ 3Hz) at $\sim$ 25V.

4.1.4   Interpretation and conclusion

The data suggests that the depleted region in the intrinsic semiconductor, i.e. the active area of the photodiode is almost linearily correlated to the applied

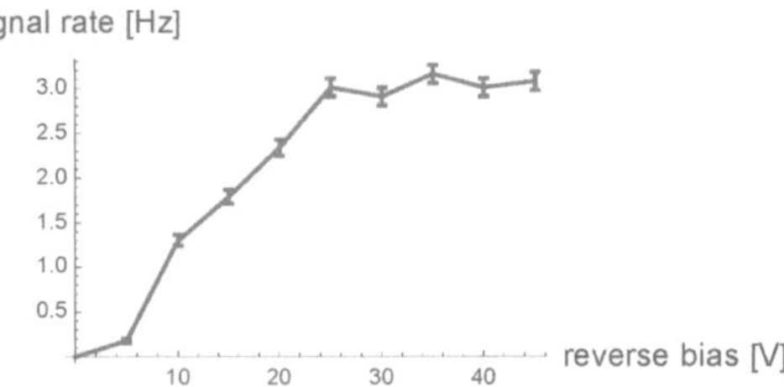

Figure 4: Plot of signal rate (corrected for background) against reverse bias voltage

reverse bias voltage, and that for voltages of 25V and higher, the active area spans over the entire intrinsic semiconductor.

In order to increase the sensitivity of the photodiode it is therefore recommended to apply a reverse bias higher than the 12V which are applied in the current SRD prototype; ideally ~ 25V.

4.2 Temperature dependence of parcel sensitivity

4.2.1 Introduction

Apart from finding the ideal reverse bias, another important question concerning the planned test flight of the Small Radiation Detector is how its parcel sensitivity changes with temperature. This is important because as the detector, attached on a weather balloon, rises in the atmosphere it will experience a substantial drop in temperature in the environment.

4.2.2 Experimental set-up

In order to investigate the effects of low temperature on the parcel's sensivity, an experiment very similar to the one described in section 4.1.2 was conducted. Only this time, the parcel was not lying on a table but rather on top of a metal plate that had been cooled down with liquid nitrogen. Metal was used because it has a high thermal capacity. The metal plate was placed into a styrofoam basin in order to facilitate cooling it down with liquid nitrogen and also to thermally insulate it from the table so it would stay at an approximately constant temperature throughout the experiment. This was double-checked by measuring the temperature of the plate and the parcel before and after the experiment.

The parcel was sitting on top of a regular document wallet on top of the metal plate in order to reduce noise by insulating the parcel from the metal plate electrically (but not thermally).

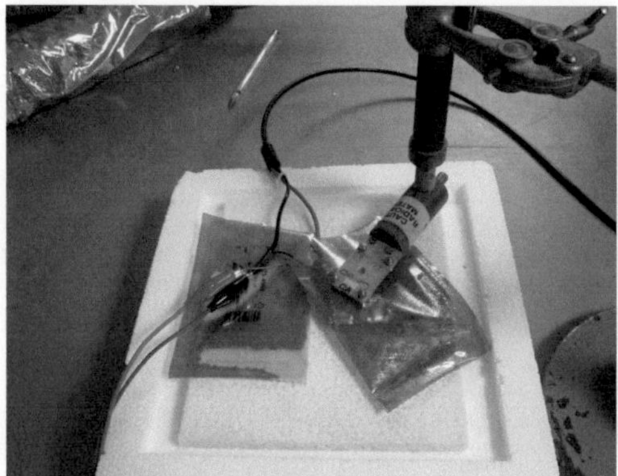

Figure 5: Early set-up of the low temperature experiment with parcel and cables inside / on top of antistatic bags and radioactive source 2U435 above the parcel

Figure 5 shows an early version of the set-up. Here, the parcel is inside an antistatic bag rather than on top of a regular document wallet. This turned out to be a less than ideal set-up because condensed water from plate crept into the antistatic bag touching the copper wrapping and thus causing noise.

4.2.3   Result and conclusion

For the experiment, the parcel was operating at a reverse bias of 12V because that would be the working voltage provided by the weather balloon in the test flight of the current prototype.

Weather balloons go up to heights of about $\sim$ 30km (mid-stratosphere), so they typically experience temperatures between $-40°$C and $+20°$C. The latter is roughly room temperature. The former could not be sustained over a long enough period of time in the lab. It was possible, however, to conduct the experiment with the metal plate at a stable $-40°$C and the parcel at a stable $-10°$C for 300s. The experiment was then repeated at room temperature with all other circumstances being exactly the same.

measurement at low temperature [parcel at $-10°$C]	measurement at room temperature
346 ± 19 counts	337 ± 18 counts

The uncertainties, again, are calculated under the assumption of Poisson distributed events. Within this uncertainty it is fair to say that there is virtually no effect of temperature on the parcel's sensitivity, at least for the temperatures examined in this experiment, which are comparable to the temperatures the parcel will be exposed to during the test flight.

5 Conclusion

After conducting the experiments described above and fixing a lot of smaller technical problems with multiple prototype versions of the Small Radiation Detector, the device was finally ready to go on a first test flight. For this purpose, it was evacuated and protected against water of any form, attached to a conventinoal weather balloon (radiosonde), and sent up into the atmosphere, as documented in figure 6. This test flight marks the end of the research project discussed in this report.
During the test flight, the detector worked as expected in the lower troposphere, but then it experienced some complications and ultimately broke down completely. Further research must be done to investigate what exactly happened.

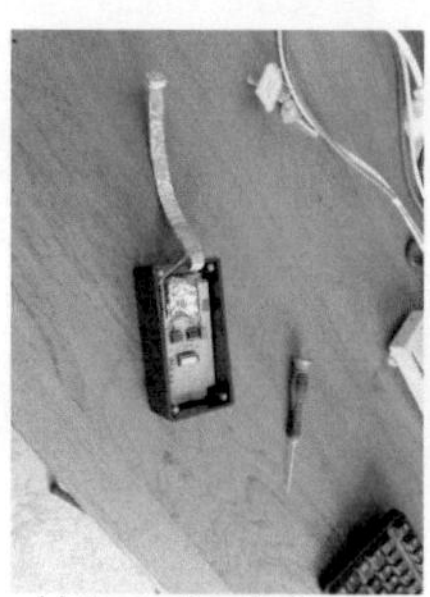

(a) Evacuating the detector

(b) Detector in box and power supply

(c) Radiosonde ready for take-off

(d) Radiosonde taking off

Figure 6: Test flight: Launch photos

A
Appendix A: Basic Specifications of the Small Radiation Detector

Parameter	Specification	Notes
Supply voltage	16V battery	Could be changed within range 4-36V. Battery powered in current weather balloon prototype. Commercial prototype could be powered from battery or phone.
Current consumption	20mA	
Sensor	$100\,\mathrm{mm}^2$ PIN photodiode PS100-7-CER-2 PIN	There is a possibility of using a smaller and cheaper sensor (or an array of them) to reduce cost and size, such as the BPW34 ($7.5\,\mathrm{mm}^2$), but as the prototype with the larger photodiode works fine we are sticking with that for now. There is also a possibility of getting some photodiodes made to our requirements, which might be economical for manufacturing.
Photodiode reverse bias voltage	12-25V	Currently working with 12V. 25V would be better (see Chapter 4.1.4).
Size	Approximately $10\,\mathrm{cm} \times 5\,\mathrm{cm}$	This was defined by the weather balloon requirements, but could be further reduced.
Mass	$\sim 30\,\mathrm{g}$	
Particles detected	Gammas, betas $>$ 350eV, does NOT detect alphas	muon detection possible, but exact pulse height and triggered counts are still to be determined
Data output	RS232/TTL to USB. ASCII format data outputs total number of counts, time since start and seconds and milliseconds, pulse height	Quite flexible. Doesn't need to be physically connected to a computer (e.g. could be Bluetooth)

Appendix A: Basic Specifications of the Small Radiation Detector (cont.)

Parameter	Specification	Notes
Pulse height detection range	TBC	We hope to be able to separate out different types of particles by their pulse height (energy), and possibly "high" and "low" energy photons, but we don't anticipate that detailed energy information will be available.
Response to a known radioactive source (count rate per Becquerel)	100% if activity is not too high	The PIC we use to detect pulses has a certain dead time which makes it miss pulses if activity $\gtrsim 0.5\,\mathrm{Bq}$. This is not a problem of the actual detector, though. It can be changed by using another PIC (software), which we did once we identified the problem, i.e. this is not a problem on the test flight.

B

Appendix B: Technical details

Measurements in Chapter 4.1

- Equipment used: Amplifier ORTEC 440A, Single Channel Analyzer Canberra 2030, for exact procedure and adjustments see section 4.1.2

- Radioactive source (2U435): ^{60}Co (half life: 5.27 a) probe bought in May 1988 with a confirmed activity of 4.40×10^5 Bq back then. $\Rightarrow$ Activity at the time of the experiment: ~ 14 Bq.

- Typical signals in this setup: Pulses with a rise time of $\lesssim 0.5\,\mu$s, pulse length of $\sim 6\,\mu$s and pulse height between 400 mV and 1500 mV.

- Measurement time: 300 s, discriminator level: 400 mV (in every run, signal and background measurements alike).

Data:

reverse bias [in V]	counts with source	counts without source
0.00	0	0
5.00	57	3
10.00	393	3
10.00	543	6
15.00	711	9
20.00	906	3
25.00	885	12
30.00	951	3
35.00	918	15
40.00	942	18

References

[1] NOAA / NSW Space Weather Prediction Center http://www.swpc.noaa.gov/info/Satellites.html

[2] Viktor Franz Hess: Über Beobachtungen der durchdringenden Strahlung bei sieben Freiballonfahrten. Physikalische Zeitschrift 13, 1912, p. 1084–1091.

[3] "Pin-Photodiode" by Kirnehkrib. Licensed under Creative Commons Attribution-Share Alike 3.0 via Wikimedia Commons - http://commons.wikimedia.org/wiki/File:Pin-Photodiode.png# mediaviewer/File:Pin-Photodiode.png

[4] S. Eidelman et al., Physics Letters B592, 1 (2004)

[5] Pfeffer, Jeremy I., Modern Physics, Imperial College Press. (2005)

[6] Karen Aplin, Monitoring of space weather and radioactivity using small airborne platforms, EGU General Assembly 2013

YOUR KNOWLEDGE HAS VALUE

- We will publish your bachelor's and
  master's thesis, essays and papers

- Your own eBook and book -
  sold worldwide in all relevant shops

- Earn money with each sale

Upload your text at www.GRIN.com
and publish for free